Generics: The Destruction of the American Pharmaceutical Industry

A Case Study in Support of Federal Term Limits

by Dennis J. Weber Ph.D

DORRANCE
PUBLISHING CO
EST. 1920
PITTSBURGH, PENNSYLVANIA 15238

Dorrance Publishing Co
585 Alpha Drive
Suite 103
Pittsburgh, PA 15238
Visit our website at *www.dorrancebookstore.com*

ISBN: 978-1-6442-6150-7
eISBN: 978-1-6442-6124-8

INTRODUCTION

This is the story of the generic drug industry, how it came to be formed, and its effect on the American pharmaceutical industry. I worked in the pharmaceutical industry for over thirty years, most of which was in research at two different companies from 1954 to 1991. I have seen the industry from the very bottom all the way up to pharmaceutical research and quality control at the Ph.D. research scientist level. I remember how it was in the 1950s and what happened in the early 1960s and the results of the damage done by Congress.

The Pure Food & Drug Act of 1906 had as its purpose the "preventing the manufacture, sale, or transportation of adulterated or misbranded or poisonous or deleterious foods, drugs, medicines, cosmetics and liquor". It also regulated traffic in medicines and for other purposes. The act expressly forbids the sale of any adulterated or misbranded medicines, either within America or to or from foreign countries. Before 1906 there were two different kinds of medicines. The patent medicines, usually sold over the counter, while the other type of medicines required a prescription from a physician. The pharmacist used his expertise to fill the prescription for the patient. The recipes for many common prescription medicines were found in the United States Pharmacopoeia (USP).

Before 1906, there was little or no regulation of the purity or safety of foods, drugs, or cosmetics. Milk was often diluted with water to the point that the white was almost gone. Whitening agents were added to hide the poor milk quality. Pills were made using the drug plus a binder and rolling the mixture by hand on a pill tile. The resulting pills were so hard that they passed through the body with no effect. For many years the Upjohn Company logo showed that such pills could be hammered into a pine board without breaking. "Buyer Beware" was the watchword of commerce.

After 1906, the government passed the Pure Food and Drug Act and formed the Food and Drug Administration (FDA). The FDA was formed to control the quality of the products sold to the public. That legislation was the beginning of the FDC specifications for food, drugs and cosmetics. In those days, there were two kinds of pharmaceutical companies. One kind made only over-the-counter products that required no prescription, the Watkins Company is an example. The other kind of pharmaceutical company sold medicines that required a prescription from a physician to be filled at a pharmacy. Pharmaceutical companies were very careful to not damage their reputation as a source of good, effective, and safe medicines because only physicians could write prescriptions and they would not write prescriptions for medicines from a company they could not trust. These times were the beginning of the American "ethical" drug companies. The word ethical meant that the products were advertised only to physicians and not to the general public. This policy allowed the physician to use his medical training to prescribe the best medicine in each situation. The physician was not badgered by untrained patients to prescribe the latest mass-advertised "miracle" medicine. The situation was better than the "buyer beware" time. The law said that the medicines must be safe for the patient but it was not necessary for the medicines to actually be effective in treating disease. This was not malicious, but only a fact of the need to grandfather those medicines on the over-the-counter market.

Pharmaceutical research was still in its infancy. The ethical drug companies' products were all safe and effective. The FDA regulations that the ethical drug companies had to meet evolved as the sciences of drug discovery and development improved. Sir Alexander Fleming in London had put a Petri dish containing a cloudy layer of germs on the open window sill. The next day he noticed that there was a cleared area in the cloudy layer around a spot where a spore from an unknown kind of organism had landed from the London air. Fleming knew that something was being produced by the strange spore that was killing the germs in the Petri dish. Fleming prepared dozens of similar Petri dishes all inoculated with a sample from the cleared area in the original Petri dish. After some years of research, the structure of the strange germ-killing substance in the Petri dish was determined. The substance was called penicillin.

By the late 1930s, penicillin was being produced in the United States. A consortium of Midwest pharmaceutical companies (Eli Lilly, Upjohn, Abbot Labs, Merck and Parke-Davis) were producing penicillin for the WWII effort to save battle field lives. That effort was the source of the Upjohn expertise in using fermentation technology to produce new medical products. In the 1930s, some sulfa antibiotics caused destruction of the kidneys and killed some patients who were given the antibiotics in too large a dose. The cause of the problem was that the sulfa drug was not soluble in the urine and when the drug precipitated in the kidney tubules, the kidney was destroyed and the patients died. This was the reason why "triple sulfa drugs" were marketed to avoid the destruction of the kidneys. Each different kind of sulfa antibiotic in the mixture had its own solubility so that the total dose of any one kind of sulfa drug did not exceed its solubility in urine. It was during this time that the pharmaceutical industry realized that more research was needed to understand the details of safety and the biological mechanisms and action of marketed medicines. The result was the formation of pharmaceutical research divisions in each of the ethical drug companies.

Steroids are another part of the increased sophistication of pharmaceutical research. Steroid hormones had been known to be biologically active for many years but there was no good source of steroids that could be obtained to use to try and develop new kinds of steroid medicines. The early 1930s was when pregnant human mothers at Borgess and Bronson hospitals were paid to allow the collection of their urine which contained measurable concentrations of steroid hormones. The hospital bill when I was born in 1934 was paid by my mother selling her urine. This was not very satisfactory, so Upjohn developed a horse farm near Richland where the urine of pregnant mares was collected. The steroid hormones from the urine was isolated and used in research to try to develop new pharmaceutical products.

Upjohn was not the only company trying to develop steroid pharmaceutical products. A Mexican company called Syntex found a large native tuber that contained appreciable amounts of one of the steroid structures. However, the tuber steroid did not have the structure necessary to be able to use it as a raw material to make medical grade steroid pharmaceutical products. There was a biologist at Upjohn who tried to find a germ that could eat and convert the unusable tuber steroid to a structure that would allow the synthesis and development of steroid pharmaceuticals. The biologist was successful and that is what started the Upjohn Company as a world leader in the manufacture of pharmaceutical products.

The formation of sophisticated pharmaceutical research was necessary to meet the requirement that any drugs, prescription or over-the-counter, sold to the public must be safe. This led to determining the toxicology of all molecules that had potential for pharmaceutical products. The drugs had to be safe but the law did not say that the drugs had to actually be effective! That is, the drug did not have to actually work! Such a law was passed in order to allow the grandfathered sale of existing over-the-counter medicines. This was the beginning of the ethical drug companies (Upjohn, Parke-Davis, Lilly, and many others) who made sure that their products not only were safe but actually worked as intended. Other so-called patent medicine companies marketed their products in a different manner.

The ethical drug companies used detail men (the term used for well-educated providers of pharmaceutical information) who called on physicians and gave the physician information about the new drugs intended to treat a disease. The use, dose, side effects, and supporting data were given directly to the physician who used the detailed information to decide when to prescribe the new drug for his patients. At the same time, this information was printed on a package insert for the physician and patient to read. All prescription medicine information was given directly to the physician via office visits. The highly-trained and medically sophisticated detail men did not sell drugs to the physician. They only informed the physician what the new product could do. The actual point of sale was only at the pharmacy under the direction of a registered pharmacist. There was no advertising to the general public. This was done so that the physician could make medical decisions about which drug product was best to use in a given situation without being badgered by parents who were ignorant of the medical details involved in the disease and the attributes of new medicines. Sales were by prescription only, which were written by the physician. The system worked extremely well. Other non-ethical drug companies, known as patent medicine manufacturers, advertised directly to the public. These were the products that most often did not actually work but were marketed to the public solely by extensive advertising in magazines, catalogs, personal opinions, and other mass media. Some of these products had names like "Carter's Little Liver Pills", "Lydia Pinkhams Women's Tonic", "Listerine", and many others.

DISCOVERY

During these times, the late 1930s, 1940s, and early 1950s, the discovery of new drug molecules followed a pattern of testing very large numbers of often randomly chosen molecules in a drug screening process. The aim was to try to find a molecule that had some kind of biological activity that might be used to treat a disease condition. Each screening process attempted to find a particular anti-disease activity. Screening molecules required a wide variety of scientific procedures. Samples were tested for antibiotic activity by measuring the effect on germ colonies in a laboratory Petri dish. Blood pressure effects in animals could be used to screen for drugs treating high blood pressure and other such medical conditions. Many other screens used other criteria that were thought to lead to useful biological activity. The screening instrumentation was primitive. Kymographs, similar to a seismograph, used smoked paper with an attached stylus to scratch a line on the smoked paper. Vitamin research used chickens to test the effect of vitamin formulations on growth and health.

When a screening procedure discovered some biological activity in the tested sample, the first thing that was done was to chemically isolate the actual molecule in the sample that was producing the observed biological activity. This is not a simple problem. Fractions of the sample had to be isolated and the fractions tested to see if a particular fraction contained the desired biological activity. Raw samples that were tested for antibiotic activity were most often just dirt obtained by random sampling from wherever seemed a likely environment. Upjohn detail men were required to send soil samples each month to be tested for possible antibiotic activity.

When the actual biologically active molecule in the sample was isolated, the real and most difficult work of drug discovery began. Physical and analytical chemistry were used to determine the three-dimensional structure of the molecule, its physical and chemical properties, and methods of analytical chemistry were developed to measure the concentrations of the molecule in various body fluids. The discovery process had barely started. Scientific publications were consulted to be sure that the biologically active molecule had not been discovered and previously reported. If the molecule had been previously reported, all efforts to develop the molecule into a marketed medicine were usually halted. This was done because a previously discovered molecule would most likely have already been developed into a product if it were possible. The fact that the molecule did not become a product usually meant that there was some problem that made further development useless. It was more rewarding to stop development on such a molecule and look for a different biologically active molecule.

Eventually, the discovery process, over many years, kept finding the same biologically active molecule and its activity, which meant that the process was becoming unproductive. This was a common problem in the search for new antibiotic compounds. If further development was justified, the next step was to synthesize a wide variety of similar but slightly different structured molecules to try and find a structure that gave a more intense anti-disease activity than the original discovered molecule. This hunt for the best structure of the molecule required synthetic organic chemists to build the various molecules and then more biological testing of each of the new molecule structures. This could take many years depending on the structural complexity of the molecule. Each of the new structure of molecules had to be tested to see if they exhibited an increase in the desired biological activity. There were no guidelines for the type of structural changes needed. The creativity and knowledge of the organic chemist was the key.

The other reason for the synthesis of these analogs of the original active molecule was to avoid a competitor's possible patent application. If a patent application became advisable, it was necessary to be sure that the claims in the patent application covered enough possible analogs of the original molecule. This was necessary so that some other company could not find an analog structure that gave an even greater desired biological response. Sometimes, the patent claims did not cover the subject well enough and a competitor company synthesized a molecule based on the original discovery with even greater medical utility. The discovery process was still years from finding a molecule worth trying to develop into a marketable product. It is critical to understand that a molecule with potential to become a marketable product was patented at the beginning of the discovery/development process. This meant that the patent clock was ticking while the potential product was still being developed. The probability was very high that the potential drug product would fail before it could be marketed. That is, it was likely to fail to pass the rigorous tests needed to be sure that the marketed product would be safe, effective, and could be sold at a price that was more competitive and better than medicines already on the market for the same disease condition. More will be said about this problem later. Suffice it to say that the remaining, useful patent life of a marketed product was extremely short, just a few years, usually less than five years. A summary of the discovery process is given here. Antibiotics, screening soil samples to find new antibiotics, plants as sources of new drug molecules. Chemical synthesis of analogs of biologically active molecules was needed. The decision step of which molecular structures to apply for a patent. At this point, the discovery process could be considered to be finished and the development process, an even longer time, would begin.

DEVELOPMENT

A summary list of the steps needed in the drug development process is given below.

[a] Toxicology in two rodents and a mammal was required by law.
[b] Determination of dose size for possible clinical trials was needed.
[c] Metabolism, excretion, absorption, and excretion balance studies in animals.
[d] Dosage form development for future human clinical studies.
[e] Pharmacokinetic and dosage regimen studies in animals.
[f] Phase one, phase two, and phase three human clinical studies.

The next step was to determine the toxicity of the drug molecule in animal models. The FDA law required this process. The pharmaceutical company had no choice but to test toxicity in established scientifically accepted methods. The law requires that the toxicity of a compound be tested in two species of rodent (usually mice and rat) and one mammalian species (usually dogs). There are several reasons for the choice of rodent and mammalian species that will be discussed later. Many months or years would pass during the toxicity testing of a potential drug product. All of the synthesized analogs that seemed active enough to have a chance to make it to market also had to be tested for toxicity. The determination of possible toxic effect on subsequent generations of test animals required many months. The design of toxicity experiments (termed protocols) tested dose size (usually measured in milligrams or micrograms of drug per kilogram of animal body weight), dosage interval (once, twice, three, etc. times a day), dosage route (oral, intravenous, subcutaneous, sublingual, transdermal, etc.), were considered and the total length of the toxicity experiments (usually measured in many months) and reproductive toxic effects of the drug molecule. Reproduction experiments required at least two or three generations of the test animal to be sure that no toxicity appeared in subsequent generations. The age of the generations tested was a critical parameter because the time required for the appearance of toxic effects was a function of animal age. This parameter required many months of continuous monitoring of the test animals. Toxicity was measured in many organ systems of the test animal. This required the preparation of highly technical microscope slides of organs that were examined under a microscope to detect any deviations from normal organ tissue. The microscope work required hundreds of slides of organ slices and the talents of highly trained scientists who could recognize when a tissue was not normal.

After the microscope work, reports were written which became part of the documentation required by the FDA for further development. The absolutely critical piece of information that the toxicity experiments were designed to provide was the dose of the drug that produced any sign of toxicity in the test animal. This dose of the drug that produced any toxic result was the starting point basis of the dose that was used to begin the calculation of the size of drug doses in the future planned human clinical studies. More will be said about the requirement for the observed toxic dose later. Because of the absolute requirement of the determination of a toxic dose in the test animal, it sometimes happened that in oral dosage of the drug it would not be efficiently absorbed into the blood and the continued dosing caused a plugging up of the rat or mouse's intestinal tract. This was a disaster because a blood-distributed toxic dose could not be determined if the molecule was not absorbed into the blood stream from the intestinal tract. These often would require injected doses and then the toxicokinetics of the resulting experiments had to be performed.

Toxicity testing was usually the graveyard of the continued development of the drug molecule. This meant that all of the years since the first observation of biological activity in the original molecule could be for naught. All the previous time and expense often yielded nothing. This was often the result in toxicity testing. If the animal toxicity testing gave results that were acceptable, the development process continued to the next set of requirements. The scientific basis for American drug development requirements were established not by the FDA or any other government agency, but by professional pharmaceutical scientists working in pharmaceutical company labs, university medical labs, chemical analytical labs, physical chemistry labs, and biological labs in America. The scientific research defined the reputation of a pharmaceutical scientist. The scientist decided what was needed, not the corporation or the government. Corporation and government bureaucracies are ignorant of the valid science required. The labs were in America because the American pharmaceutical industry was the world leader in pharmaceutical research development in those times. More will be said about this later.

The reason why larger mammalian species (dogs) were used was that timed blood samples following an oral, intravenous or suppository dosage route had to be taken. The test animal had to have a large enough blood volume so that the timed multiple samples (about two or three milliliters) over several hours or days did not adversely affect the normal functions of the test animal. The blood, urine, sweat, and etc. samples were chemically analyzed to measure the change in drug blood concentration over time. This required the development of analytical chemical procedures that were capable of measuring drug concentrations in the nanogram (one billionth of a gram) per milliliter range. Such analytical procedures require extremely sophisticated, expensive instrumentation. Gas chromatographs, high pressure liquid chromatographs, chromatography using mass spectra as a detector, extremely sensitive biological analytical methods, and nuclear magnetic resonance instruments were all part of the tools that were used to develop analytical procedures to measure very small concentrations of the drug molecule in animal and human body fluids.

Animal drug metabolism studies (balance studies) measured the percent of an oral or intravenous dose of the drug that was eliminated by various routes. It is important to know if a drug is completely eliminated from the body and by which routes. If it is not completely eliminated, the fraction remaining in the body must be organ located so that the data can be used to hunt for the same data in humans. Urine, feces, sweat, saliva, etc. were all collected and the drug and its metabolites were measured and compared with the dose given to determine if all or only some fraction of the dose was eliminated from the body.

Pharmacokinetics is the science of the absorption, distribution, metabolism and excretion (ADME) of a drug from the body. The data is used to determine the rate at which the drug is eliminated from the body. The pharmacokinetic data is used to calculate the proper dose interval, dose size, and dosage route that will give the proper concentration of drug in the human blood stream sufficient to treat the disease condition. A blood concentration can be too low (insufficient to stop the disease) or too high (blood concentrations may be in the toxic range). Pharmacokinetic studies are first done in animals, usually dogs, and then in humans when the clinical studies are allowed to begin. These experiments use sophisticated drug analytical methods developed by the scientist and very sophisticated mathematical analysis of the time dependence of the drug concentration data.

While these preclinical animal studies are running, research pharmacists develop and test dosage forms (tablets, intravenous solutions, capsules, suppositories, etc.) that will be needed for the human clinical studies. These dosage form experiments may take months to finish. If the sum total of the preclinical animal data suggests that the drug molecule will not be acceptable, the whole project is terminated and all the time and expense of drug discovery, toxicity, and pharmacokinetic analysis is lost. It may take more than a year for all the preclinical animal studies to be finished.

If the data is acceptable, an Investigational New Drug Application (INDA) is filed with the FDA. Many months may pass while the FDA scientists inspect all the preclinical data to see if the drug can be allowed to be tested in humans. If the FDA is satisfied that the preclinical animal data is acceptable, the INDA will be approved. If the INDA is not approved, the scientists go back to the preclinical animal studies to run more experiments to try to satisfy the FDA objections.

CLINICAL TESTING

At this point, when an INDA has been approved by the FDA in the discovery and development of a drug product, the drug molecule may be given to humans for the very first time. The usual way to run clinical studies is to start with a safety study [Phase 1] to evaluate the toxicity in humans. Phase 2 tests the response of the target disease in humans to the experimental, possible new drug. The number of diseased people in the Phase 2 study is kept to about a dozen because of the difficulty of finding diseased people with the target disease. Phase 3 clinical studies use much larger numbers of volunteers and the study is often aimed at evaluating possible interactions with concomitant dosed patients.

Pharmaceutical companies often have numerous drug candidates in the product pipeline and so must make decisions which drug candidates are most important and most likely to progress to a marketable product. The resources are limited with regards to the number of patients available that have the target disease. One study [Hoos, W.A. et al. J. Clin. Oncology, v. 31, pg.3432-3438, 2013.] showed that when several pancreatic cancer clinical trials are done in parallel, they compete for a limited pool of patients. For example, the study referenced above would require participation of 83% of patients with surgically treatable tumors. Only about 5% of those are willing to volunteer to participate in a study.

Pharmaceutical human clinical studies are divided into three phases. Clinical Studies Phase One is a safety study. The purpose is to detect any toxicity that was not seen in the animal toxicity studies. The animal toxic dose that was mentioned before is divided by one hundred or more as the starting dose in normal humans. That is, the first dose in normal humans is done very, very carefully and at an extremely small dose. This dose in normal human volunteers is slowly increased, using a standard protocol, until some sign of biological response or toxicity is detected. These studies are done in normal human volunteers. This sign may be as simple and as harmless as a rise in temperature, some itching, change in blood pressure, other vital signs, etc.

The data from the study are analyzed statistically to determine the validity and importance of any sign of toxicity. When any sign of toxicity is detected, the study is halted and from then on, no human will ever again receive that dose. If a suicidal person takes a huge dose to kill himself, that dose can exceed the toxic dose. The critical question that the drug safety study is designed to answer is, "Is the dose that produced the very first toxic response too small to have any chance to cure the disease that the drug is designed to treat?" If the dose is considered too small to treat the disease, the whole project is dropped and all the years, efforts, and expense are lost.

All the years and expense of the preclinical work are at risk now. The questions to be asked and answered are:

(1) Is the drug toxicity found in humans sufficient to stop the development of the medicine?
(2) Is there any dose or route of administration of the drug that will give a blood concentration in humans sufficient to cure the disease without simultaneously causing any toxic reaction?

In other words, is it possible to give the drug in a dose that will cure the disease and not cause toxic side effects? Does the drug interfere or be interfered by drugs that may already be in the body for other medical reasons? These are called drug interaction studies. These questions are the basis for Phase 1, Phase 2 and Phase 3 clinical studies in humans. The Phase 1 studies have been described above. The Phase 2 studies are the first studies where the drug is given to humans that actually have the disease that is being treated. This study is designed to answer the question—is the dose that will treat and alleviate the disease less than the toxic dose found in the Phase 1 safety study. If the second clinical study finds that there is no or insufficient desired biological activity at a dose less than the toxic dose found in the Phase One clinical study, then the whole effort is abandoned and the pharmaceutical company has to start all over again in the discovery of a different drug molecule and everything has to be repeated for the next possible new medical molecule. There is no guarantee that even at this late stage in drug research the drug will actually help to alleviate the targeted disease. This disaster is not unknown. The Upjohn Co had a new drug that avoided much of the brain damage caused by stroke. The Phase 2 clinical study showed it worked well as expected in male patients at the required dose but did not work as needed in women at that dose. The effective dose in women was too high and so the whole project, after about twelve years, was dropped.

Phase 3 clinical studies are much expanded from the Phase 2 studies and incorporate many more patients that are also taking other medicines for their disease or other disease problems. These are the drug interaction studies. The medical tools used to evaluate interaction studies include MRI, x-ray, chemistry of body fluids, concentrations of the new drug, and the metabolites of the new drug or metabolites of other medicines that the patient needs to take, and vital signs (blood pressure, pulse rate, etc). The stage 1, 2, and 3 human clinical studies are extremely expensive and require many years to find enough patients who have the targeted disease and then collect enough data to answer the clinical questions. It is very difficult to find enough patients who have the necessary medical conditions to adequately answer the clinical question being asked.

By the time human clinical studies are completed and acceptable at least ten to thirteen years have passed. The New Drug Application (NDA) data are reviewed by the FDA. This can take a few years to adequately review the mountain of data submitted. If the FDA approves the NDA, then permission is given to the pharmaceutical company to market the new medicine. This is called an Approved New Drug Application (ANDA). Sometimes there is literally a truck load of copies of the data sent to the FDA. The twelve to thirteen years or more necessary for discovery, toxicity, development, and clinical testing represent most of the years of patent life that are lost and can never be regained.

There is no guarantee that the human clinical studies will give data that will allow the company to file an NDA (New Drug Application) with the FDA. Several more years may pass while the FDA scientists analyze the clinical data to see if the drug is safe, effective, and does not dangerously interfere with other drugs that may have been given at the same time. The NDA may be accepted or it may be refused. If it is refused, all the years, money, and effort are wasted.

The FDA has done a good job in many cases to protect American medicine from potentially dangerous medicines. A case in point is the European drug thalidomide. Thalidomide treats morning sickness in pregnant women. When the German manufacturer wanted to sell the drug in America, the FDA reviewer found the data troubling and refused to give them permission to market thalidomide in America. Later it was discovered that thalidomide, when given at a critical time in gestation, caused anatomical birth defects. Some American women got the drug illegally and had infants with serious birth defects. The FDA reviewer did her job correctly.

Because humans are not all the same biologically, the testing is not done yet. Even after the new drug is on the market, the testing continues. After the new medicine is on the market and millions of patients, perhaps all over the world, have received the new medicine, the results are reviewed (this is called Phase 4 testing) and if there has been an unforeseen effect that is unacceptable, the new medicine is removed from the market and all the twelve to fifteen years of work is down the drain. Because not all humans are the same, it sometimes happens that certain races do not respond to the medicine as expected. There is a drug, isoniazid, used to treat tuberculosis that acts differently in Western human populations than in Asian populations. The Asians metabolize isoniazid at a faster rate, so the correct dose in Western humans is much different than the correct (larger) dose in Asians.

The Senator Estes Kefauver Commission that held hearings in the late 1950s and into 1960 on drugs, both prescription and over-the-counter, wrote and had passed into law legislation that said that all prescription and non-prescription drugs must actually be effective for the claims that were advertised to the public and to physicians by the manufacturer. Up until this time, the FDA required the manufacturer only to prove that their products were safe and not toxic. That is, they did not have to actually work! However, the ethical drug manufacturers such as Upjohn, Merck, Lilly, Squibb Parke Davis, etc. did the very expensive research to prove that their products were effective and spent tens (or hundreds) of millions of research dollars to prove it. The research to prove efficacy was done to keep a reputation with the prescribing physicians for safe and effective products. The safety of medicines was mandated by Congress in the 1930s due to the toxicity of some sulfa drug products sold at that time. The drug safety laws were very good laws as were the drug efficacy laws passed by the Kefauver Commission.

DISCLAIMER

There were some unintended consequences of the drug efficacy laws. Some of us may remember Carters Little Liver Pills, Lydia Pinkhams Tonic, Dr. Rose's Obesity Powder, Dr. Worden's Female pills, and the long list of unproven claims that Listerine antiseptic carried on the bottle label and other such non-prescription products sold by Watkins and some other companies. I made a survey of over-the-counter pharmaceutical products at a local drug store to see what were listed as "active ingredients" and what were the claims listed on the product label. The question I asked was, "Was the requirement by the Kefauver Commission that all over-the-counter pharmaceutical products be effective for the claims on the package label?"

Listerine antiseptic was one of the few over-the-counter products to survive the purging of products that were not effective for the claims on the label. Listerine survived the 1960s testing by removing all claims that could not be proven to be true. The present Listerine product claim is that it used only as a topical antiseptic (that is, it kills many germs on the skin). The Listerine package contains the following information that confirms the effectiveness of Listerine antiseptic: "The ADA Council on Scientific Affairs acceptance of Listerine Antiseptic is based on the finding that the product is effective in helping to prevent and reduce gingivitis and plaque above the gumline when used as directed."

Other over-the-counter products do not have the FDA or ADA Council approval on the package. Instead, many over-the-counter products have the following statement: "This statement (label claim) has not been evaluated by the FDA and is not intended to diagnose, treat, or cure any disease or prevent any disease." This disclaimer means that the law requiring all drugs to be effective is now a dead letter as far as many over-the-counter products are concerned. In many cases, we are back to the situation that existed before the Kefauver laws were passed. Except now, we have the generic market and most prescription drug prices are drastically higher. Experience is no virtue in government. Incompetents need to be voted out of office and federal term limits need to be instituted.

Results of Inspection of Present Over-the-Counter Pharmaceutical Products

---

PRODUCT NAME
INGREDIENTS
LABEL CLAIMS

PRICE

1. Watkins Liniment
- Active Ingredient
- 3.5% Camphor
- (Counter irritant)
- (no disclaimer!)
- Backache, Arthritis, Strains, Sprains, Bruises, Temporary Aches, Pains of Muscles and Joints
- Not Shown

2. Heart Burn MD
- Proprietary Blend
- (This means that the manufacturer will not say what is in the product)
- Treats Heart Burn
- $16.99

3. Cholest-Off Sodium
- Plant Sterols 900 mg
- Pantethne 150 mg
- (antihyperlipoproteinmic)
- [Statement - This statement (label claim) has not been evaluated by FDA and is not intended to diagnose, treat, cure any disease or prevent any disease]
- Lowers Cholesterol Naturally
- $34.99

4. Triple Flex
- [The statement (label claim) is not intended to diagnose, treat or cure any disease or prevent any disease.]
- Treats Joint Pain
- $24.99

5. [Store name] SAM-E
- S-Adensoylmethionine 400 mg [Anti-inflammatory treats chronic liver disease]
- [Statement -The statement (label claim) has not been evaluated by the FDA and is not intended to diagnose, treat, cure any disease or prevent any disease.
- Mood & Emotional Support, Joint Comfort, Liver Health
- $29.99

6. Breathe Right

- Water, alcohol, glycerin Polysorbate 80, essential oils
- Snore Relief
- $???????

7. Cooling Blue Ice

- Menthol 2%
- [Anti-itch, anti-pruritus[itch]
- [NB - pruritus is the medical term for an itch]
- Pain Relief
- $4.99

8. Joint Flex

- Camphor 3.1%
- [Anti-itch treatment-not pain relief]
- Arthritic Pain Relief
- $18.99

9. Capzasin HP

- Capsaicin 0.1%
- [counter irritant]
- Arthritic Pain Relief
- $9.99

10. Bengay

- Camphor 4% [anti-itch]
- Menthol 10% [anti-itch]
- Methyl Salicylate 30% [anti-itch]
- [Caution-ingestion of small amounts can be lethal]

- Pain Relief
- $8.99

11. Blue Star Ointment

- Camphor 1.24% [anti-itch]
- Powerful Pain Relief
- $7.49

12. Aspercreme

- Trolamine Salicylate 10% [anti-itch]
- Caution ingestion of small amounts can be lethal
- Pain Killer
- $6.79

13. Australian Dream

- Histamine Dihydrochloride 0.025%
- [gastric secretion, anti-allergenic]
- Pain Killer
- $29.99

14. Absorhine Jr.

- Menthol 1.27% [anti-itch]
- Pain Killer
- $7.49

15. Memoryl

- Vitamin E, Folic Acid
- [Statement - The statement (label claim) has not been evaluated by the FDA and is not intended to diagnose, treat or cure any

disease or prevent any disease]
- Note—This product was imported from Italy]
- Memory Improvement
- $39.99

16. Pre-Tense
- Combination of plant and other natural substances
- [Plants and Substances used not stated]
- [Statement - This statement has not been evaluated by FDA and is not intended to diagnose, treat or cure any disease or prevent any disease.
- Relieves Nervous Tension
- $24.99

17. Relaxane
- Combination of plant and other Natural substances [Plants and Substances not stated]
- [Statement—the statement (label claim) has not been evaluated by the FDA and is not intended to diagnose, treat or cure any disease or prevent any disease.
- Stress Relief
- $10.99

18. Finest Natural
- Proprietary Blend (note this means that the manufacturer will not say what is in the product) [Statement -The statement (label claim) has not been evaluated by the FDA and is not intended to diagnose, treat or cure any disease or prevent any disease.
- ?????

- Not Stated!
- $8.99

19. Joint MD

- Mixture of plant and other Natural substances
- [Statement - The statement (label claim) has not been evaluated by the FDA and is not intended to diagnose, treat or cure any disease or prevent any disease.
- Joint Stiffness and Cartilage Support
- $14.99

Eight of the nineteen products are antipruritus (anti-itch) formulations. The products are expensive to just fix an itch. These formulations work by de-sensitizing the skin nerve endings so that the itch is not felt so much. However, none of the formulations will correct the cause of the itch. They mask the itch, but any real medical problem is left hidden. The Merck Manual of Diagnosis and Therapy states that pruritus can be a symptom of skin or systemic diseases. The above over-the-counter products hide the symptom but do nothing to diagnose the cause of the pruritus.

The disclaimer on many of the products tells us that many current over-the-counter products are no better than the products that were on the shelves of drugstores before the Kefauver laws were passed. To a large extent, we are no better now than we were before the efficacy laws were passed in the early 1960s. However, now the American pharmaceutical industry has been damaged and prices have risen to troublesome values. It is important to always read the label on any over-the-counter medicine. If the disclaimer statement is present, do not buy the product. If the claims are not supported by the label data, be very careful what you buy.

It must be stated that much of the price rise in medicines is due to the lower purchasing power of the dollar. When the government spends money that it does not have, the Federal Reserve Bank loans, at interest, trillions of dollars, the value of the dollar and what a dollar can buy decreases. Taxpayers and the American economy are being damaged by our own government! Besides requiring ethical drug manufacturers' (Upjohn Co., Parke-Davis, Merck, Lilly Squibb, etc.) prescription drugs to be effective, the Kefauver laws also required that over-the-counter drugs be effective.

Under the leadership of the Food and Drug Administration, the United States pharmacopoeia (USP), numerous schools of pharmacy, the National Formulary, and others devoted the next ten years to analyzing all the over-the-counter drugs for content and efficacy to meet the requirements of the Kefauver laws. The analyses were done in chemical and biological laboratories under the guidance of the USP, an independent organization separate from the manufacturers and the FDA. Those products whose claims were not valid were told to eliminate the false claim(s), to do the research necessary to prove their claims, or to take the product off of the market. It was during this time that Carter's Little Liver Pills and others disappeared from the drugstore shelf and the claims on the Listerine antiseptic bottle were severely reduced. Those companies whose products did not meet the efficacy requirements, and who were not willing to spend the millions of dollars to research and develop new products, decided to go out of business.

If this were the end of the story, the American consumer, the ethical drug manufacturers, and the medical community would be in better condition today and the cost of drug prescriptions wouldn't be the political football it has become. The drug companies who were put out of business hired lawyers to lobby Congress and began asking for a way to sell medicines without having to spend any money for research to prove safety or efficacy claims. They even appealed to Congress with the story that ethical drug manufacturers were making "obscene" profits and should be forced to give, free of charge, the results of years of expensive pharmaceutical and medical research to anyone who asked for it. The costs of pharmaceutical, medical research were borne one hundred percent by the ethical manufacturers themselves.

The government spent nothing for the discovery and development of new drugs, with one exception. In the case of cancer, the National Cancer Institute screened new molecules for anticancer activity and, if a promising candidate was found, it was offered to the ethical drug manufacturers for further research on anti-cancer activity, toxicity, dosage form development, and clinical testing. The ethical drug manufacturers who chose to try to bring these anti-cancer drugs to market bore the total cost of development of the potential cancer drug. The ethical drug companies bid against each other for the right to develop the potential anti-cancer drug. The United States government recouped part, if not all, of the cost of finding the potential cancer drug.

The General Accounting Office in July 2003 was the latest federal agency to refute the claim that the cost of drug development is borne by the government. In year 2001, the government had licensing rights to only six of the top one hundred brand-name drug products purchased by the Department of Veterans Affairs and to only four of the top one hundred brand-name drugs that the Department of Defense dispensed. That is, only six of the top one hundred drug products had ANY federal funds in their product development cost. These were mostly anti-cancer drugs.

The politicians could not resist the votes and re-election possibilities and passed laws giving FREE all of the extremely expensive research results and complete recipes to make the drug product to the "generic" companies as they came to be called. The "obscene profits" that the lawyers' and lobbyists loved to talk about were nothing but smoke and mirrors. The patent life of a drug was seventeen years in those days. However, what the politicians and lawyers ignored and hid from the public was that the patent clock starts ticking when the MOLECULE is discovered NOT WHEN THE PRODUCT GOES TO MARKET. This requires at least twelve to fifteen years before scientific proof of safety and efficacy and dose development is completed and the FDA allows the drug product on the market.

The development time for a marketed drug in those days WAS MORE THAN TEN YEARS. Today, it is even longer because the government regulations make it even more difficult! Alan F. Holmer in the August 19, 2003 issue of Insight Magazine points to a development time period of ten to fifteen years. Holmer, a spokesman for the Pharmaceutical Research and Manufacturers Association, also gives data on the fraction of drug molecules that successfully make it to market. ONLY ONE OF five thousand screened molecules is approved by the FDA as a new medicine. Of these five thousand compounds, two hundred and fifty enter preclinical testing and development and only five of those make it to human clinical testing and of the five in clinical testing, only one is approved by the FDA and makes it to market. Only three of every ten marketed drugs generate enough sales revenue to match or exceed the cost of research and development of that drug! The two hundred and fifty preclinical and five clinical failures ALL represent money down the drain that must be balanced by the sales of the products that make to market.

The cost of preclinical and especially clinical testing of a new drug is enormous. The total average cost of developing and bringing to market a new pharmaceutical product, according to the Tufts Center for the Study of Drug Development, was $802 million in year 2000, up from $318 million in 1987. The cost in 1965 was less than one-third of the 1987 cost. These costs include the expense of research failures and the impact that long product development times have on investment costs. The time left in the patent life of a newly marketed product was less than five years! Those few years' profits had to pay for the tens or hundreds of millions of dollars spent by the drug company in discovery, toxicity testing, dose development, clinical testing, and marketing costs of the new product.

Also, the lawyers never said that the fraction of molecules that were patented and made it to market was MUCH LESS THAN FIVE PERCENT. That is, over ninety-five percent of the cost of research by drug companies was on failed medicines and profits from those drugs that made it to market had to pay the cost of ALL the research on both successful and unsuccessful drug candidates. In spite of all this, the United States ethical drug industry was so good at developing new drugs that they could still stay in business. The United States had the best medicines anywhere in the world! There was a reason why the system worked, but the reason was never mentioned to the voting public by the lawyers or politicians. While bringing a drug to market was extremely expensive, only those new drugs that were better than the drugs already on the market for a particular disease could be expected to successfully compete in the market place of physicians, writing prescriptions and thus give a return on investment. For this reason, when a patent life ran out, other ethical drug companies chose NOT to make and try to market a copy of the innovator's product because the copy still had to have data gathered by the copier, proving the copy product had safety and efficacy and still had to sell it at a price competitive with the same drug product that was already on the market for the same disease.

For this reason, the huge cost of developing new drugs and paying the cost of failed research could be recouped because it was usual for the innovator to enjoy selling the product for years UNTIL A COMPETITOR MARKETED A DRUG THAT WAS BETTER THAN THE ONE ALREADY ON THE MARKET. The system guaranteed that only better medicines would make it to market and the public would then enjoy the very best pharmaceutical product possible. The physicians knew that a new prescription medicine was better than what was already on the market for a particular disease. All this was destroyed in the 1960s by the generic drug laws. The ethical drug manufacturers warned Congress what would happen to the industry, but politicians chose to ignore the warning and looked only at votes for re-election. Laws were passed in the 1960s forcing the ethical, innovator drug companies to give, free of charge, all the information necessary to make any product that the generic company decided to ask for when the patent life on the original active molecule ran out. The ONLY requirement asked of the generic drug company was that they make several batches of the medicine product and submit data to the FDA demonstrating that the generic product batches FUNCTIONED the same as the original innovator's product. The word "functioned" means that the generic copy of the drug was absorbed, distributed, metabolized, and eliminated in the body the same as the innovator's product and that the drug content of the dosage form was the same as the original marketed drug. MANY OF THE GENERIC DRUG COMPANIES CONTRACTED OUT THE LABORATORY WORK NEEDED TO PROVE COMPLIANCE. Even so, some of the generic companies were unable to reproduce the product even when they had all the recipes needed to make it. Some of those failed generic companies went to a drug store, bought the innovator's product, and submitted it as if it were their own work. Generic companies who resorted to this deception and were discovered were usually put out of business.

Some generic drugs to this day do not reproduce the original marketed product but are still allowed to be sold. Dilantin (anti-seizure medicine) is one such drug that has unacceptable generic competition. I have personal experience of the dangers of using invalid generic forms of Dilantin products. It is common knowledge by most physicians and pharmacists that the generic copy of Dilantin does not work as the original product works. The hundreds of millions of dollars spent by the innovator pharmaceutical company to bring a new product to market must be recovered by sale of the new medicine in order for the recovery of the cost of the research to be recovered. The initial cost to the patient of a new, better medicine product must be high enough to recover the hundreds of millions of dollars spent in research to bring the product to market. After some years, about five years or more, when the sale of the new product has recovered the research cost, the price to the patients is dropped so that any other drug company that is considering bring a new medicine to market for the same disease problem will have to produce a new medicine for the same disease that is significantly better than what is already on the market.

If a generic product is allowed to be sold, it becomes impossible to recover the hundreds of millions of dollars spent for the scientific research needed for FDA approval to sell the new medicine. If it is impossible to recover research costs then the research for new medicines stops. All the high salary science research jobs are lost and the market for medical research jobs is destroyed. Thus, Congress has damaged the economy by destroying the market for medical research jobs. That is exactly what happened to the Upjohn Company and now all medical research jobs are gone from Kalamazoo, Michigan. The lower cost manufacturing jobs remain but the high cost of medical research not recovered destroys the market for the research jobs.

Congress, in order to gain votes, has destroyed the market for medical research jobs. There are several other job markets that are destroyed by Congress in the hunt for votes. Government welfare with no labor attached destroys the market of the need for workers in an industry. The government keeps raising the minimum wage which destroys the market for new businesses able to be created. The American economy is damaged! There are many dozens of television advertisements by lawyers who want to bring lawsuits against companies who are selling defective products because adequate research was not done.

Before generic drugs were allowed to be sold, long term toxicity data were required to avoid side effects. Of course, the lawyer culture wants as many side effects as possible. The physician is forced to exactly specify in the prescription which product is to be dispensed in order to avoid the generic product. My own personal experience shows that insurance companies ignore the problem and still try to require the generic form of Dilantin be dispensed instead of the innovator's original product to increase the insurance companies' profit when Congress in the middle 1960s forced the innovator pharmaceutical company to give, free of any charge, all the recipes, methods, or any other process needed to make the innovator's product.

CONSEQUENCES

The American public is not getting the quality of drugs it used to get. Buying drug products from foreign companies gives no guarantee that the product is what it's purported to be. Some foreign purchased drugs have turned out to be nothing but water! It is also common for the foreign manufactured drugs to contain less of the biologically active drug than is needed for therapeutic efficacy. It is not possible for the FDA to analyze every batch of prescription drugs imported into America. The foreign countries have no need to maintain a physician's trust for excellence in their products. The foreign manufacturers have no need and the United States cannot check efficacy of all imported medicines. The result of the advent of generic drug products is that the ethical drug companies no longer can be assured of recouping the cost of research and development of new drug products. I was a research scientist at the Upjohn Co. when all this was developing. There were only a few choices available to the ethical drug companies. The financial choices that ethical drug companies had to make were:

(a) go out of business
(b) look for a buyer or a merger partner to try to stay in business
(c) stop expensive research on new drugs and sell only the ones existing in their catalog, which eventually were obsolete
(d) decide that the company's research capabilities were good enough to still discover, develop, manufacture, and market

medical products and compete with the generic and other ethical drug companies

The Upjohn Company thought they were capable of competing. Upjohn attempted to use the prostaglandin family of drug molecules along with other drugs designed for the treatment of stroke as the next sources of new products. They were wrong, but they did try and cannot be blamed for trying.

The Upjohn name does not exist anymore in medical products. Upjohn, Syntex, Squibb, Dow Pharmaceutical, Parke-Davis, Warner-Lambert, and many other famous drug companies do not exist anymore or are just part of larger organizations. Foreign drug companies who do not have to give their products to generic companies do not have the problem are buying American companies and in some cases exporting jobs to offshore countries. Research expertise, if it even exists, is cheaper in other countries. Greed and political expediency are destroying the American pharmaceutical industry, reducing the quality of medical care, exporting jobs to other countries and putting the United States on a slippery slope to national medical mediocrity. Those ethical drug companies who are still in business (Merck, Lilly, Pfizer, etc.) are forced to raise the price of their products in order to pay for the research and development of new drug products that is the heart of their existence. Merck's Executive Retention Plan Offers Severance says, "Merck, seeking to keep managers and other company leaders from jumping ship after the $38 billion loss in market value since September 30, 2004, have been forced to adopt a plan to give severance packages to two hundred and thirty executives should control of the company change and become a take-over target. Merck faces thousands of lawsuits from lawyer's clients who claim Vioxx hurt them or their relatives." The lawyers who have advocated suing the company smell money and are willing to destroy a pharmaceutical company to get it. This is one more step in the destruction of the American pharmaceutical industry. Merck's market value is about 61.4 billion, which is half that of Glaxo-SmithKline, an English corporation, and less than one-third of Pfizer, the world's largest pharmaceutical company. First it was the tobacco industry, then the target became the fast food industry and now the target is the pharmaceutical industry. Lawyers have no conscience or integrity! Their self-respect is based solely on their client's bottom line.

When the price goes up, the generic companies, who do no research, seize the opportunity and raise their prices also. Generic companies currently lead the pack in the percentage rate of increase in drug prices! This history of generic drugs is not new. It was known as far back as the 1970s and the current state of affairs was predicted in the 1960s when Congress was considering the generic market. The ignorant, uninformed, government-does-it-all, socialist attitude of the politicians, lawyers, and media is part of the problem.

Can the mistakes that were made be corrected and the price of prescriptions brought back to more reasonable levels? I don't think so. The generic drug manufacturers have a huge financial stake in the industry and besides, the ethical drug manufacturers, in order to meet the generic competition, must maintain prices. The biggest losers are the American public and the patients who die because the medicine that could have healed them could not be brought to market. If you believe that government laboratories could do the research and development and bring new drugs to market, you have not understood what the problem really is. The present research efforts of the ethical, non-generic pharmaceutical companies are focused on bioengineered drug products. These products are extremely difficult to produce and may eventually cause the generic drug industry to decide that it would be too expensive to make the raw materials needed in such drug products.

The generic companies can always send the jobs to other countries to try to meet the FDA requirements.

However, Congress might force the innovators of the bio-engineered medical products to sell the critical ingredient to the generic company so that the generics too will continue to damage the American ethical pharmaceutical industry. This is a sad state of affairs. What can we do to avoid such mistakes in the future? I believe the greed and thirst for power and re-election can only be controlled by removing the possibility of making a lifelong career out of politics. That means federal term limits. When the United States Congress passed laws forming the generic drug industry in the 1960s, the existing pharmaceutical industry had warned Congress that the laws would destroy the American pharmaceutical industry.

By the early 1970s, the effect of the laws was apparent to the ethical drug manufacturers and a strategy was needed to counter the loss of the market if the American drug manufacturers were to stay in business. The term "ethical drug manufacturer" means that physicians were visited by drug detail men whose function was to meet with the physician and give the scientific medical details why the product was better for a particular disease than what was already on the market. The real advantages of the new drug were emphasized and the physician could write prescriptions for the new product if he was convinced that it was better than what was already on the market. The physician decided which drugs to use and the insurance companies stayed out of the loop. It used to be that insurance companies were most troubling to physicians, NOT ANY MORE! It is government regulations passed by medically ignorant bureaucrats sitting at a desk that are now the most troubling.

There was no advertising in general distribution magazines allowed by mutual agreement of the American Pharmaceutical Manufacturers Association (PMA). These days, magazine advertisements aimed at mothers and other people who do not understand the details of medical treatment have produced overuse of antibiotics which have lost the ability to kill germs that were once easily removed from the body. People are dying from infections that were once easily cured. Those manufacturers that had opted to continue pharmaceutical research needed some way to recover the part of the market lost to the generic companies. Most of the companies were forced by the realities of the market place to form their own generic subsidiaries and essentially "join the enemy". This strategy allowed the ethical, research-based drug companies to take advantage of the generic drug laws passed in the 1960s. Cross-licensing of marketed products among the research-based manufacturers became the method of competing with the generics. Cross-licensing allowed a company to increase its product list without the expense of research to develop new products. Each cross-licensed product was marketed with a different trademarked name that the ethical drug manufacturer chose. This meant that many different trademarked medicines were really the same product. This required enormous increases in advertising costs if the cross-licensed product had any chance to compete with the equally good product on the market from the innovator or other companies that had cross-licensed the same product for their own catalog.

The politicians and lobbyists loved to point at the increased advertising expenditures as a reason for federal control of the manufacturers. This, when Congress itself had caused the problem in the first place in the 1960s by passing laws forming the generic drug industry. Those generic companies who successfully manufactured a copy-product still faced a problem.

"Struggling New Jersey pharmaceutical company Schering-Plough Corp. and German chemical and drug maker Bayer AG have formed an alliance under which Schering-Plough will take over U.S. marketing of Bayer's primary-care medicines. The partnership, announced Monday, September 13, 2004 could give Bayer a bigger presence in the world's biggest drug market, while offering Schering-Plough badly needed new products to sell and a potentially key marketing partner for some of its drugs in Europe and Japan. Under the alliance, Kenilworth-based Schering-Plough's 3,500 salespeople (pharmaceutical detail men) will take over U.S. marketing of Cipro and Avelox, both broad-spectrum antibiotics for respiratory and skin infections, plus the blood-pressure drug Adalat and some small established primary-care drugs. Schering-Plough's sophisticated detail men sales force also will handle U.S. promotion of Bayer's impotence treatment Levitra. 'Schering-Plough needs products and this is a way of getting them temporarily until they can develop their own.' said drug-industry analyst Mike Krensavage of Raymond James & Associates. 'Bayer certainly could use the marketing assistance of Schering-Plough.' Schering -Plough said the partnership, expected to become effective on Oct. 1, would slightly reduce earnings for the rest of 2004, partly due to integration and transition costs, then would add slightly (italics mine) to earnings starting in 2005."

The primary-care products will remain the property of Bayer and continue to be sold under the Bayer brand name. This means that Schering-Plough will not be classed as a generic drug producer and that most of the profits will go to Bayer, a foreign drug manufacturer. The research jobs will remain in Germany and only the sales jobs will be filled by American citizens. "Schering-Plough will pay a 'substantial royalty' to the German company on net sales of the products," Schering-Plough spokesman Steve Galpin Jr. said. Bayer is best known in the United States for its aspirin.

The pharmaceutical industry advertised only to physicians via the detail men. The generic companies did not have any such detail men expertise and could not compete unless the laws were changed to allow direct advertising to the general public. The direct advertising put the physician out of the drug information loop and resulted in uneducated people badgering the physician to prescribe this new "wonder drug I read about in *Good Housekeeping* magazine". In reality, the "wonder drug" was not new, not a wonder at all. It was just a copy of what was already on the market under a different name. This reminds me of the man who murdered his parents and then pleaded for mercy from the court because he was an orphan. Congress caused the problem and then complained that it existed. At the same time, the patent life clock for a product was ticking and when the patent expired, the generic companies jumped in and began competing with the innovator as well as those companies who had previously licensed the medicine from the innovator manufacturer. The end result was that NO ONE gained anything, the research on new medical products was greatly reduced, and the physicians and public had fewer choices and fewer new drug products were added to the market.

People died who did not ever get the chance to be treated with drugs that might otherwise have been brought to market. The huge increases in advertising costs required to compete meant higher selling prices to the public, higher costs to the insurance companies, and eventually federal and state laws to control the runaway market costs. The politicians love to pose as saviors of the consumer. Just to be sure about the actions of Congress in the 1960s I went to the congressional records at the library at Western Michigan University here in Kalamazoo, Michigan. WMU has a complete set of the congressional records available.

Entries on Jan/10/1966 to October 22, 1966. Search words are:

- PMA,
- Drug IND.
- Sales Decrease causes crisis.

- A3685
- Remarks in Senate
- drug prices 26405
- pricing practices 2866, 2889
- Remarks in House pg. A3685, 3686
- Hon. Abraham J. Molter
- fakers and swindlers in field of health have become big business.
- Let buyer beware
- Vitamins - American diet gives enough vitamins
- Arthritis remedies, cancer, cosmetic medicines, "miracle chemicals" and "miracle scientific discoveries"
- Remarks in Senate
- Sen. Long of Louisiana

> (note, the cost of discovery, development, clinical testing, registration, and marketing is ignored in these prices. The government is considering only the manufacturing cost.)

If a person needs a tetracycline, we would pay him so much for it. Tetracycline is a wonder drug that one can buy in a great number of places for five cents per pill. It costs one and a half cents to manufacture.

We are willing to pay five cents for each of sixteen pills

Every bacterium would be killed by the time a person took all sixteen capsules!

But if one wants to buy PanAlba [Pfizer] = tetracycline. Costs thirty cents a pill, used to cost fifty cents a pill.

Our government can buy same pack of two pills for two cents a pill. We provide for soldiers at two cents a pill. Pills are provided to Congressmen at two cents per pill

Drug store prices = thirty cents or fifty cents per pill for Squibb products. Achromycin = thirty cents per pill. Government can buy same pill for one and a half cents and we are willing to pay five cents per pill.

Mr. Murphy

I can answer from my experience as a business executive that if this were done by a business firm [not government] it would be referred to as unfair business competition.

We would tell the manufacturer what we would pay for it and the government would be in the position of being the biggest customer. The government could be charged with price fixing. However, if we want to run the cost of this thing [government buys medicine] up to five hundred million all we have to do is let the drug companies have their say about it.

a. Senator Long of Louisiana

My father at one time was a patent medicine salesman. He had two medicines. One was named "High PoplarLowRum" and the other was "Low PoplarHiRum". Both bottles were the same size one sold for fifty cents and the other for a dollar. High PoplarLowRum was made from bark skinned down the tree Low PoplarHiRum was made from bark skinned up the tree.

MARKETS

I t is the total cost of discovery, development, clinical testing, registration, and marketing and manufacturing that must be considered when evaluating the price of a medicine. As was shown in the above text, it was only the manufacturing cost that the ignorant politicians thought was needed to be taken into consideration. The following data and calculations show what the result is when only the manufacturing cost is considered.

Assume the total cost to discover, develop, clinical testing, registration, marketing, and manufacture a medicine is 100 million dollars. This is a very conservative number.

100 million dollars = 100,000,000. = 1.0×10^8 dollars.

The manufacturing cost [which was the only cost considered by the politicians] is 6 cents per tablet dose = 0.06 dollars = 6×10^{-2} dollars.

1.0×10^8 dollars divided by 6×10^{-2} = 0.166×10^{10} = 1.66×10^9

1.66×10^9 is the number of doses that must be sold at 6 cents per dose to just break even with the total cost of developing the new medicine.

The United States population in the 2010 census was 308 million people. 308 million people = 308,000,000. people = 3.08×10^8 people

1.66×10^9 doses divided by 3.08×10^8 people = 5.38 doses for every person in America

BUT not every person in America has the disease that is to be treated. So, if the number of people in America that have the disease is thirty million people i.e. one in ten people have the disease which would be a huge, disastrous medical problem. Then each person with that disease must buy and ingest 53.8 doses [5.38 doses divided by 1/10 of population = 53.8 doses] just for the pharmaceutical company to break even for the discovery, development, clinical, registration, marketing, and manufacturing costs of 100 million dollars to bring that medicine to market. Thus, this cost to the patient would be 53.8 doses x 0.06 dollars cost per dose = 3.228 dollars per patient. But, if the number of people who have the disease is only three million people [which is still in the range of an epidemic] = 3 million people = 3.0 x 10^6 people

1.66 x 10^9 doses divided by 3.0 x 10^6 people with the disease then 1.66 x 10^9 divided by 3.0 x 10^6 = 0.553 x 10^3 – 553 doses just to break even with the cost of bringing the medicine to market. 553 doses multiplied by 0.06 dollars per dose = $33.18 cost just to break even with the total cost of bringing the new medicine to market.

FIRST BOTTOM LINE

It is the size of the market for that disease that determines the number of doses that can and must be sold to just break even with the total cost of bringing that new medicine to market. If the cost to bring a new medicine to market is 200 million dollars [not an unreasonable number] then, the size of the market needed must double or the price per dose must double in order to just break even with the total cost of bringing the new medicine to market.

Market = 30 million doses. Then must sell 2 x 53.8 doses = 107.6 doses just to break even with total cost. 107.6 doses are 107.6 x 0.06 dollars per dose = $6.45 per dose.

If the Market – 3 million doses, [more realistic] then must sell 1,106 [one thousand one hundred doses] just to break even with cost of bringing the new medicine to market. 1,106 doses x 0.06 dollars per dose = $66.36 cost to the patient.

SECOND BOTTOM LINE

I t is the size of the market for the particular disease being considered and the total cost [total cost = total research cost plus manufacturing cost] that determines the price of a medicine plus the number of doses needed to be sold to treat the disease.

Note - there are two situations that need to be considered. Is the disease a chronic disease that may require years of treatment or is the disease an acute disease which requires fewer doses? The fewer doses needed to treat the disease, the higher the price of the dose required to just break even with the cost of bringing the new medicine to market.

Those ethical drug companies who survived did so by buying up the competition to obtain their products. Buying the company who was the innovator of a highly prescript product was more cost effective than trying to compete using the generic subsidiary licensing strategy. Thus, big companies became ever larger, the number of research-based companies continue to decrease, and foreign companies buy the American companies to obtain their products and then jobs are sent overseas to lower-cost labor markets and less government regulation interference.

The August 2004 issue of *Scientific American* (pg. 24) reports the transfer of drug research and development and clinical testing out of the United States to India. The jobs are leaving the United States because the ethical drug manufacturers must look for a way to compete with the generic marketed products. Pharmaceutical giants Novartis, Pfizer, and Eli Lilly have commissioned Indian firms to perform R & D services, including clinical testing. As "foreign companies set up shop in India, foreign expertise will grow. Each year business has grown sixty to eighty percent with almost ninety percent coming from international (American?) companies."

Before 1970, American pharmaceutical companies who wanted to market their products in foreign countries were usually faced with a special set of research and clinical studies mandated by the targeted foreign countries' governments. For the privilege and permission to market the product in the foreign country, the American pharmaceutical companies were required to repeat some research studies already completed in the United States. These studies were often part of the drug research and clinical phase of the approval process. Most of the countries, since they did not have the equivalent expertise of the United States' FDA, depended on FDA approval for marketing in the United States as proof of the medicine's safety and efficacy. These additional studies were required by the host country to give their economy a boost and to obtain a part share of the profits made by the American pharmaceutical company. This was a reasonable request in most cases. The American pharmaceutical companies viewed these redundant studies as an additional but scientifically unnecessary cost to marketing in the foreign country. The data obtained from the additional studies played no critical part in the decision to allow the drug product to be marketed in the host country. The host country still relied on the American FDA approval of the drug product for safety and efficacy assurance.

This distinction is important because it shows that the American pharmaceutical industry did not need to reduce research costs by exporting the studies to prove safety and efficacy, but did so only as an exercise to satisfy the host countries' expectation of a share in the profits. However, after the advent of Congress' institution of the generic drug industry and the concomitant loss of funds to support research on new drugs, the American pharmaceutical industry began to export critically needed studies.

EXPORTATION OF JOBS

The pharmaceutical industry looked for countries whose research expertise was sufficient to satisfy the American FDA requirements to prove safety and efficacy. Over the last thirty years these countries have developed their expertise and now jobs that used to be available to American scientists are now exported to other countries. Predictably, Congress and the media scream about the "greed and lack of concern for American jobs" when in truth, Congress' greed and lust for power caused the problem. High paying, tax revenue producing, American jobs are leaving the United States. Congress was warned in the 1960s that this could happen, but they paid attention only to the next election and the number of votes they could get to stay in power. It is a lie that re-election and political experience produces better congressional decisions. The profits that the surviving pharmaceutical companies make are a function of reduced numbers of American research efforts, reduced research building, and the use of low labor costs in foreign countries. It is ignorance and the egocentric power of government that is destroying America. The government now wants an increase in federal employees to better monitor drugs imported from other countries, all of which would not be necessary if Congress would stop trying to fix something that was not broken forty-seven years ago. I am glad I am seventy-three years old. I will be dead before my grand and great-grand kids have to live in an America destroyed by our own government.

Chairman John Dingell (D - Mich.), head of the U.S. House Committee on Energy and Commerce, and Rep. Bart Stupak (D - Mich.) chairman of Energy and Commerce's Subcommittee on Oversight and Investigations, have an alarming issue that touches the health of all American individuals and families. The issue is the U.S. government's systemic failure to inspect foreign pharmaceutical manufacturers. In the past year Americans have learned that "globalization" appears to mean allowing unsafe foreign made products into the American market. For example, there have been numerous recalls of unsafe toys. The country that is most guilty of exporting dangerous goods to the United States is China, but the Chinese are not alone.

As if that weren't bad enough, recent press accounts reveal that the U.S. Food and Drug Administration (FDA) has a major crisis on its hands regarding the inspection of foreign drug manufacturing plants. One catastrophe in particular has captured the public's attention. A Chinese plant that produced the main ingredient for the anti-coagulant Heparin was never inspected. (NB: the molecule in Heparin was first designed to be used as a rat poison!). The imported drug has been linked to twenty-one U.S. deaths and hundreds of cases involving complications. The dose of heparin in humans that avoids the poison effect must be closely monitored. The Chinese did not put the correct amount of "rat killer" in their exported product.

Reports have also surfaced about a Chinese company that manufactures Methotrexate, a drug that is used in the worldwide cancer and abortion trade. These news stories have brought attention to a November 2007 report by the American Government Accountability Office (GAO-08-224T) that shows clearly how the inspection system for imported drugs is in a state of meltdown. Only two percent of Chinese plants are inspected annually and from 2001-2007 only twelve percent of Chinese plants had been inspected compared with ninety percent for both Ireland and Switzerland. We (American Family Research Council) have also learned that the FDA inspects American plants every two years, but no required inspection period exists for foreign plants. In the United States the FDA conducts unannounced inspections of manufacturing facilities, but it does not do this overseas. Furthermore, the FDA does not supply translators to its inspection teams, so they typically must rely on translators provided by the business being inspected. The American Family Research Council believes that legislation should be enacted to protect the health of our people. Any new laws written to address this problem should include the following provisions:

The FDA must specifically hire and train staff, including translators who understand the terms of pharmaceutical manufacturing processes to conduct foreign inspections. The FDA must rapidly work toward a goal of inspecting foreign plants at rates equal to those for American plants, approximately once every two years.

Americans should be informed when they buy drugs as to the country of origin of the drug product and ingredients they are purchasing.

"Country of Origin" information should be made available on the FDA website.

Products made in foreign plants that are not subject to unannounced inspections should not be importable into the United States

Congress should authorize and appropriate sufficient funding to carry out these objectives. Because many countries will not allow inspection of their pharmaceutical manufacturing plants by American inspectors, the inspections listed above can probably not be done.

This emphasizes just one of the consequences of Congress' original mistake in passing laws during the 1960s that resulted in the generic drug trade and the destruction of the quality of American pharmaceutical products. The United States Government has caused the exportation of American pharmaceutical research to other countries. The maze of regulations makes it very difficult to stay in business in America.

When Medicare was first passed into law in the middle 1960s, the price of prescription drug products was so low that Congress did not find it necessary to include any prescription drug benefit in the Medicare laws. It was only after Congress screwed-up the process that the drug prices spiraled out of control and are still rising. When President Bush signed the recent Medicare Prescription Drug Benefit Bill, it put even more pressure on the American pharmaceutical industry to survive and now the new law itself has become a super-expensive boondoggle that threatens the United States budget and is making the voters confused, angry, and will raise their taxes to a level never before seen! It is only thirty years later, after the American pharmaceutical industry has been damaged, that the cost of prescription medicines has become a problem.

SOLUTION?

The only way to correct the damage done by Congress is to return the chance for an innovator medicine producer to make profits that will defray the cost of drug research in the United States. There is a way to do this. There are two possible solutions to the problem of the destruction of the American pharmaceutical industry. The high cost of drugs can be brought down if the law that destroyed the ability of ethical drug companies to recoup their research costs can be changed. The critical point in the process is the length of time that the drug company that did the research is allowed to SELL a product that contains the new drug molecule. Presently, the patent clock starts when the molecule is discovered so that only about five years of exclusive patent life is left when (and if) the new drug product goes to market. The length of the exclusive marketing time must be returned to the pharmaceutical company that paid for the research to bring the new product to market.

There are two possible ways to accomplish this. The marketing patent life clock should start when the product is first put on the market. This allows at least seventeen to nineteen years, the present length of a drug patent, for the drug company to recoup its investment in research and development. As it stands now, the patent clock starts when the biologically active molecule is discovered, not when it is first marketed. However, the protection of the intellectual property of the company who first discovered the molecule must be maintained so that the same molecule could not be developed by drug companies who did not spend any funds to discover the molecule. This requirement would need a change in how the patent process functions and probably would not be a politically easy change. The change would return the profits to the company who did the research to bring the new drug molecule to market. The generic market would not be able to copy the drug until the marketing patent had run out.

The other possible solution to the drug cost problem is to return the system to what existed before Congress screwed it up in the 1960s. As mentioned before, the previous system worked because it made no sense to copy a marketed drug because it would be no better than what was already on the market. Thus, the funds not spent on advertising to outsell the same drug sold under a different trade name could be saved and put to use in discovering new drugs. As long as there was no product on the market better than what was already there, the physician-based, not magazine-based, advertising worked well. A possible way to return to the system that worked is to prevent any generic copy of a drug to be sold as long as the drug in question was the best drug for that particular disease. The physicians know what the best drug is and prescribe accordingly. When a better drug product for that disease was brought to market, then the generics would be allowed to copy and sell the second-best drug for that disease condition.

I know the politicians and public would scream that the best drug is being withheld from the public for profit reasons, but the alternative is to destroy the American pharmaceutical industry, reduce the number of new drugs being brought to market, force sick people to forgo any cure, and continue the transfer of high paying technical jobs to foreign countries. Ronald Reagan was correct, government is not the solution to a problem, government is the problem! All that is needed is to allow the generic industry to copy a patent-expired prescription drug only after a better drug than the patent-expired drug has gained FDA permission to be marketed. That is, the generic companies have permission to market the "second-best drug". This will give the innovator ethical drug companies a chance to recoup the cost of research and to pay for research on new medicines without sending jobs to foreign countries.

Isn't it time the American voter said "enough is enough" and demand federal term limits to keep politicians from making decisions based on power and re-election? The philosophy of the political party that puts the welfare of the American citizen first will win the trust of the voters and will gain the opportunity to make the United States a better place to live.

www.ingramcontent.com/pod-product-compliance
Lightning Source LLC
Chambersburg PA
CBHW061518180526
45171CB00001B/236